Division Workbook Grade 3

Practical Math Exercises

DAN STEWART

TABLE OF CONTENTS

Grouping the objects as stated 5

Divide into equal groups 7

Division Facts 9

Write the division factor of given number 11

Divide by repeated subtraction 13

Division of 2-digit numbers 16

Fill in the blanks 17

Solve the division and write the other division fact for same equation 18

Division with remainder 19

Division by 10, 100, 1000 with remainder 20

Solve the following using Long Division Method 21

Solve the following using Long Division Method with remainder 24

Division solve and check 26

Shade the boxes in the given grid as per equation and then solve it 29

Divide by skipping numbers on number line 30

Divide to complete 33

Division Matrix 36

Word Problems 39

Match the following 42

Solve the division and then match the division with its other division fact 44

Solve the following division problems 45

Write down all the division and multiplication facts of given numbers 49

Division Puzzle 52

Grouping the objects as stated

There are 9 candies. Circle the 3 candies at a time. Write equation of division and solve:

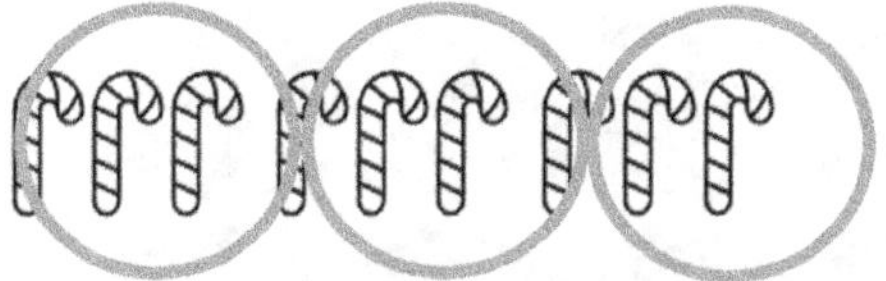

$$9 \div 3 = 3$$

I have 12 apples, divide them into group of 4 apples. Write equation of division and solve:

There are 15 pencils. Place 5 pencils in each box. How many boxes will be there?

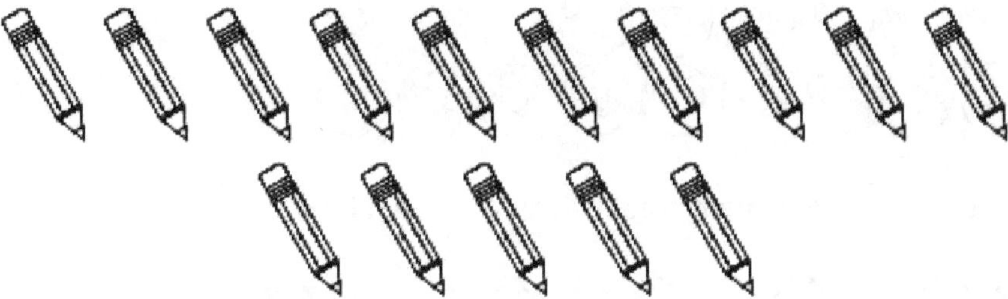

Count the total balls and place 6 balls in a box. Write the equation of division and solve:

There are 10 toys. Give 2 toys to each child. How many child will get toys?

Grouping the objects as stated

Pack 6 books in one box. How many box will you require to pack 42 books?

Pack 4 hair ribbons in one packet. How many packets will be required to pack 16 ribbons?

Serve 2 donuts to each of the guest. How many guests will be served by 26 donuts?

There are 48 eggs. Place 8 eggs in each tray. How many total trays will we require?

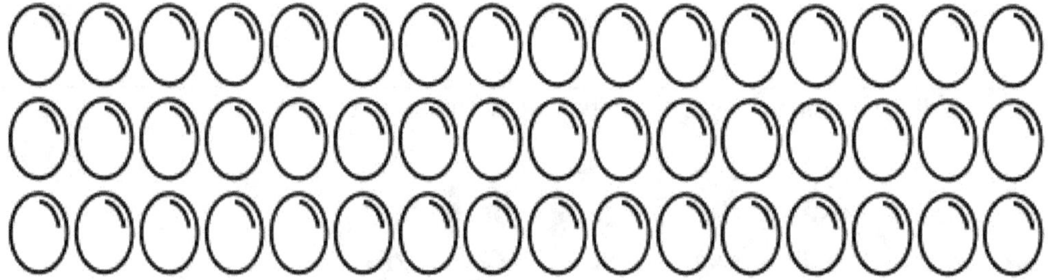

Count and Divide into group of 4 each:

Divide into equal groups

Count and divide into 5 equal groups. Write the division equation and solve:

Divide equal number of candies to 10 students. Write the division equation as well:

Divide the stars in to 4 groups in such a way that each group contain equal number of stars:

Monkeys are going to play basketball. Divide them into 2 equal size team:

Divide the following animals of the forest into 3 teams:

Divide into equal groups

Divide the following coins into 3 children of Mr. John. Each child should get equal number of coins.

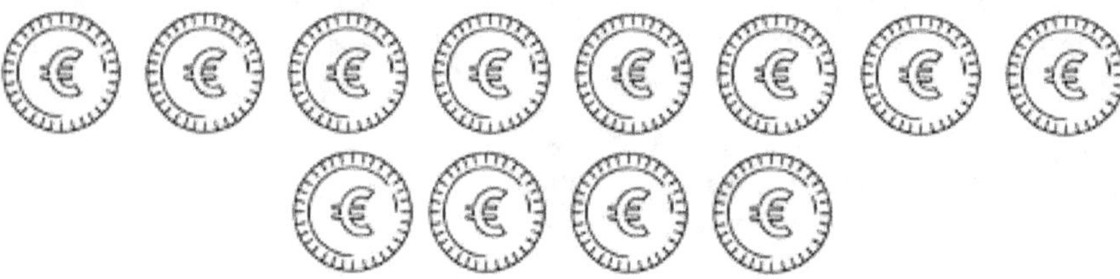

Divide these 27 people into 3 groups. Each group should have equal number of persons.

Divide 15 balls in to 3 equal size groups:

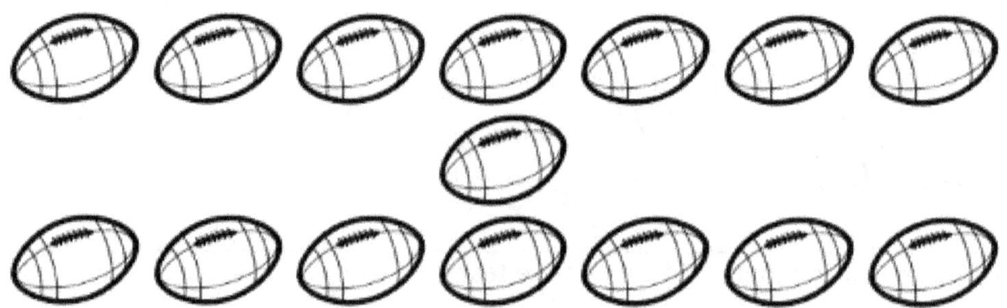

Distribute these notebooks among 7 students. Each student should get same number of notebooks.

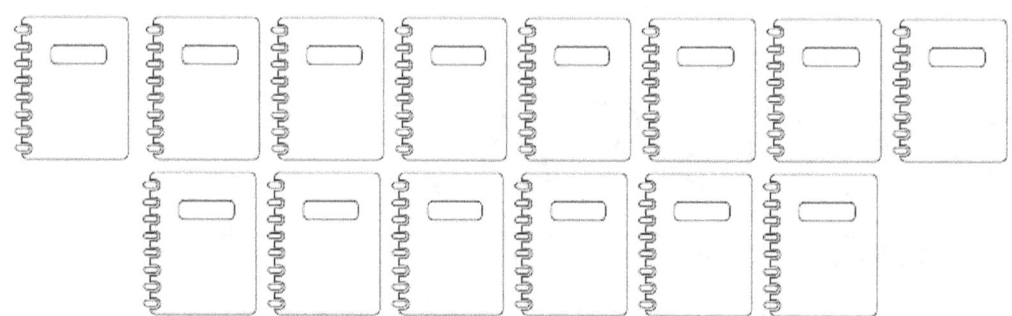

Distribute these ginger bread among 2 kids. Each should get same number of breads.

Division Facts

Here are 12 dots, which can be grouped in two ways:

This can be written as 12 ÷ 4   This can be written as 12 ÷ 3

12 ÷ 4 and 12 ÷ 3 are called division facts

Now divide the following in two ways and write its division facts

Division Fact 1: _______________   Division Fact 2: _______________

Division Fact 1: _______________   Division Fact 2: _______________

Division Fact 1: _______________   Division Fact 2: _______________

Division Fact 1: _______________   Division Fact 2: _______________

Division Facts

Division Fact 1: ______________

Division Fact 2: ______________

Division Fact 1: ______________

Division Fact 2: ______________

Division Fact 1: ______________

Division Fact 2: ______________

Division Fact 1: ______________

Division Fact 2: ______________

Division Fact 1: ______________

Division Fact 2: ______________

Write the division factor of given number

100 ÷						
2	4	5	10	20	25	50
= 50	= 25	= 20	= 10	= 5	= 4	= 2

80 ÷						
2	4	5	8	10	20	40

96 ÷						
2	3	4	6	8	12	16

120 ÷						
2	3	4	5	8	10	12

120 ÷						
15	20	30	40	60	1	120

Write the division factor of given number

150 ÷						
2	3	5	10	15	25	50

429 ÷						
1	3	11	13	33	39	143

525 ÷						
3	5	7	15	21	25	35

650 ÷						
5	10	13	25	26	50	65

1000 ÷						
2	4	5	8	10	20	25

Divide by repeated subtraction

Example: **30 ÷ 5** 30 − 5 = 25 ----- 1 25 − 5 = 20 ----- 2 20 − 5 = 15 ----- 3 15 − 5 = 10 ----- 4 10 − 5 = 5 ----- 5 5 − 5 = 0 ----- 6 **30 ÷ 5 = 6**	24 ÷ 4	18 ÷ 6
28 ÷ 7	32 ÷ 8	42 ÷ 7
15 ÷ 3	18 ÷ 3	16 ÷ 4
36 ÷ 9	40 ÷ 5	45 ÷ 9

Divide by repeated subtraction

15 ÷ 5	24 ÷ 6	60 ÷ 10
56 ÷ 8	48 ÷ 6	45 ÷ 5
42 ÷ 7	64 ÷ 8	63 ÷ 9
65 ÷ 5	68 ÷ 17	52 ÷ 13

Divide by repeated subtraction

30 ÷ 3	60 ÷ 12	75 ÷ 15
81 ÷ 9	90 ÷ 9	96 ÷ 16
91 ÷ 7	112 ÷ 14	108 ÷ 8
105 ÷ 35	150 ÷ 30	175 ÷ 25

Division of 2-digit numbers

54 ÷ 6 =	24 ÷ 6 =	42 ÷ 2 =	45 x 9 =
80 ÷ 10 =	84 ÷ 7 =	30 ÷ 6 =	80 ÷ 5 =
80 ÷ 8 =	60 ÷ 3 =	84 ÷ 3 =	38 ÷ 1 =
63 ÷ 3 =	10 ÷ 5 =	48 ÷ 2 =	72 ÷ 6 =
36 ÷ 3 =	60 ÷ 3 =	60 ÷ 1 =	57 ÷ 1 =
27 ÷ 9 =	56 ÷ 7 =	24 ÷ 8 =	25 ÷ 5 =
24 ÷ 4 =	88 ÷ 2 =	63 ÷ 7 =	75 ÷ 5 =
45 ÷ 9 =	66 ÷ 3 =	84 ÷ 6 =	66 ÷ 2 =
57 ÷ 3 =	9 ÷ 9 =	81 ÷ 9 =	36 ÷ 4 =
18 ÷ 6 =	26 ÷ 2 =	32 ÷ 4 =	48 ÷ 6 =
33 ÷ 11 =	56 ÷ 7 =	64 ÷ 8 =	96 ÷ 5 =
93 ÷ 3 =	85 ÷ 5 =	90 ÷ 9 =	80 ÷ 4 =
94 ÷ 2 =	35 ÷ 7 =	66 ÷ 3 =	22 ÷ 2 =

Fill in the blanks

___ ÷ 17 = 6	104 ÷ ___ = 8	180 ÷ ___ = 10	126 x 14 = ___
84 ÷ 12 = ___	14 ÷ 7 = ___	___ ÷ 2 = 4	___ ÷ 19 = 6
12 ÷ ___ = 3	___ ÷ 13 = 6	54 ÷ ___ = 9	84 ÷ ___ = 7
___ ÷ 6 = 7	78 ÷ 13 = ___	24 ÷ 6 = ___	120 ÷ ___ = 6
171 ÷ ___ = 9	30 ÷ ___ = 3	___ ÷ 16 = 7	11 ÷ ___ = 1
24 ÷ ___ = 6	56 ÷ 8 = ___	___ ÷ 4 = 12	___ ÷ 19 = 10
___ ÷ 11 = 3	18 ÷ ___ = 3	48 ÷ ___ = 4	___ ÷ 11 = 7
30 ÷ ___ = 6	60 ÷ ___ = 10	26 ÷ 13 = ___	___ ÷ 7 = 9
45 ÷ 15 = ___	___ ÷ 8 = 5	___ ÷ 13 = 8	54 ÷ ___ = 9
14 ÷ 14 = ___	___ ÷ 11 = 5	16 ÷ ___ = 16	___ ÷ 7 = 4
60 ÷ ___ = 4	35 ÷ 5 = ___	81 ÷ ___ = 9	50 ÷ ___ = 5
___ ÷ 6 = 8	___ ÷ 10 = 15	27 ÷ ___ = 3	12 ÷ 4 = ___
___ ÷ 7 = 7	55 ÷ ___ = 55	108 ÷ ___ = 9	10 ÷ 1 = ___

Solve the division and write the other division fact for same equation

Division Fact	Other Division Fact		Division Fact	Other Division Fact
15 ÷ 5 = 3	15 ÷ 3 = 5		350 ÷ 50 =	
145 ÷ 5 =			231 ÷ 11 =	
99 ÷ 9 =			391 ÷ 17 =	
108 ÷ 9 =			187 ÷ 11 =	
135 ÷ 3 =			660 ÷ 55 =	
75 ÷ 5 =			240 ÷ 12 =	
88 ÷ 8 =			156 ÷ 12 =	
60 ÷ 3 =			100 ÷ 25 =	
135 ÷ 5 =			459 ÷ 27 =	
140 ÷ 10			620 ÷ 20 =	

Division with remainder

42 ÷ 5 = 8 remainder 2	117 ÷ 12 =	690 ÷ 70 =
174 ÷13 =	170 ÷ 86 =	67 ÷ 67 =
749 ÷ 32 =	577 ÷ 76 =	580 ÷ 41 =
1016 ÷ 51 =	697 ÷ 55 =	440 ÷ 88 =
950 ÷ 77 =	797 ÷ 92 =	794 ÷ 41 =
187 ÷ 98 =	972 ÷ 77 =	801 ÷ 31 =
550 ÷ 85 =	125 ÷ 7 =	282 ÷ 34 =
906 ÷ 22 =	431 ÷ 64 =	241 ÷ 18 =
419 ÷ 5 =	583 ÷ 39 =	146 ÷ 73 =
608 ÷ 93 =	770 ÷ 48 =	551 ÷ 64 =
259 ÷ 40 =	248 ÷ 66 =	841 ÷ 70 =

Division by 10, 100, 1000 with remainder

63 ÷ 10 = 6 remainder 3	85 ÷ 10 =	71 ÷ 10 =
147 ÷ 10 =	106 ÷ 10 =	277 ÷ 100 =
349 ÷ 10 =	757 ÷ 100 =	1650 ÷ 100 =
1100 ÷ 100 =	1690 ÷ 10 =	2044 ÷ 1000 =
420 ÷ 100 =	970 ÷ 100 =	94 ÷ 100 =
718 ÷ 100 =	1002 ÷ 1000 =	3002 ÷ 100 =
5500 ÷ 10 =	1501 ÷ 1000 =	2009 ÷ 100 =
609 ÷ 1000 =	13 ÷ 100 =	214 ÷ 100 =
194 ÷ 100 =	350 ÷ 10 =	614 ÷ 10 =
806 ÷ 100 =	707 ÷ 100 =	5115 ÷ 1000 =
2950 ÷ 1000 =	2844 ÷ 10 =	4810 ÷ 10 =

Solve the following using Long Division Method

2 ⟌ 50	3 ⟌ 84	2 ⟌ 18
2 ⟌ 40	6 ⟌ 84	3 ⟌ 42
6 ⟌ 78	8 ⟌ 56	2 ⟌ 44
5 ⟌ 65	7 ⟌ 42	8 ⟌ 24

Solve the following using Long Division Method

10	240		79	474		30	300

66	132		44	616		18	396

13	637		11	132		93	465

95	380		21	840		20	360

Solve the following using Long Division Method

48) 2400	63) 8883	77) 2618
14) 9674	36) 8604	91) 9009
26) 9438	73) 3212	94) 1222
52) 2288	50) 4650	49) 3675

Solve the following using Long Division Method with remainder

97) 796		61) 240		58) 729

89) 300		28) 953		23) 676

79) 762		80) 643		95) 623

28) 748		81) 725		22) 201

Solve the following using Long Division Method with remainder

32 ⟌ 3596	60 ⟌ 6714	65 ⟌ 7912
84 ⟌ 6217	96 ⟌ 7615	91 ⟌ 4312
61 ⟌ 5747	68 ⟌ 7431	40 ⟌ 6771
47 ⟌ 7844	55 ⟌ 1212	17 ⟌ 9108

Division solve and check

Solve:

Dividend ÷ Divisor = Quotient + Remainder

98 ÷ 16 = 6 remainder 2

Check:

(Quotient x Divisor) + Remainder = Dividend

(6 x 16) + 2

= (96) + 2

= 98

Now solve these and check:

Solve	Check
35 ÷ 11 =	
19 ÷ 16 =	
69 ÷ 16 =	
52 ÷ 10 =	
39 ÷ 25 =	
11 ÷ 11 =	
26 ÷ 18 =	

Division solve and check

Solve	Check
21 ÷ 10 =	
59 ÷ 17 =	
65 ÷ 11 =	
12 ÷ 15 =	
39 ÷ 12 =	
20 ÷ 34 =	
26 ÷ 15 =	
35 ÷ 10 =	
79 ÷ 28 =	
83 ÷ 13 =	
71 ÷ 20 =	

Division solve and check

Solve	Check
298 ÷ 97 =	
781 ÷ 69 =	
856 ÷ 93 =	
488 ÷ 61 =	
362 ÷ 92 =	
641 ÷ 84 =	
203 ÷ 74 =	
370 ÷ 76 =	
391 ÷ 77 =	
997 ÷ 69 =	
822 ÷ 13 =	

Shade the boxes in the given grid as per equation and then solve it

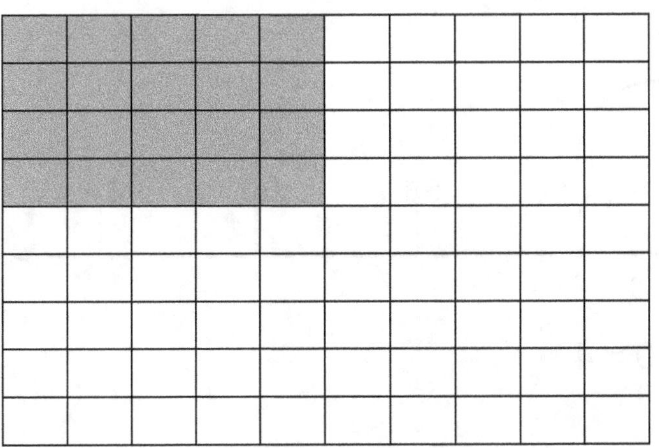

$20 \div 4 = 5$

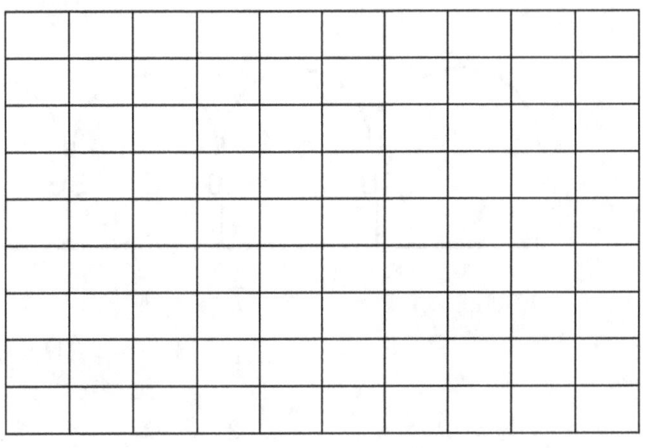

$35 \div 5 =$ ___

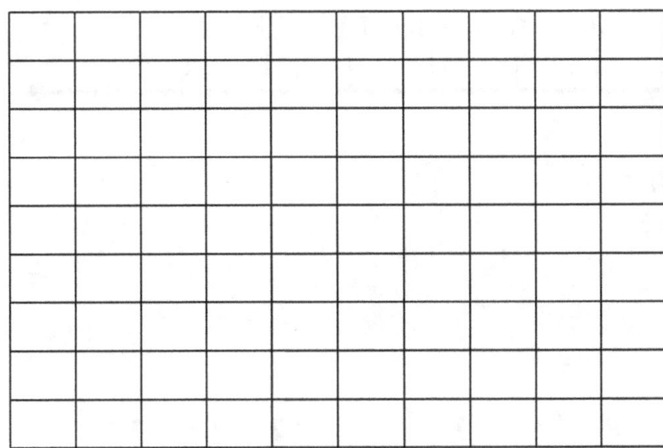

$28 \div 7 =$ ___

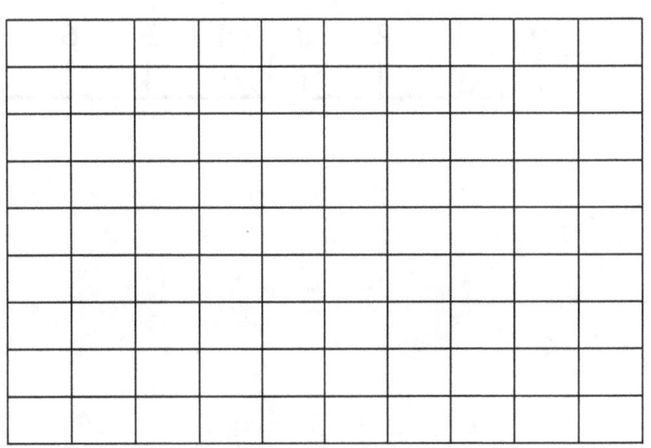

$42 \div 7 =$ ___

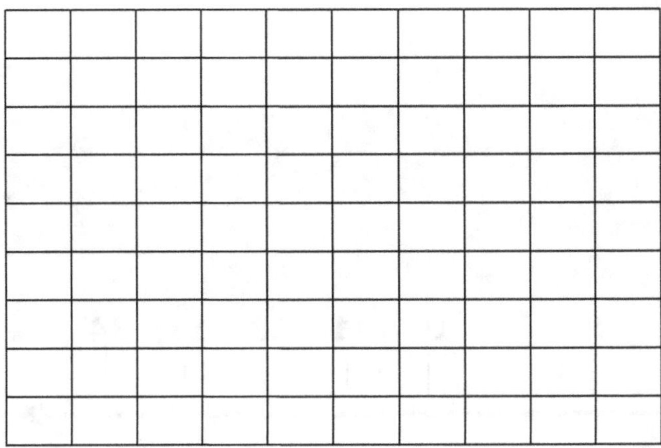

$36 \div 6 =$ ___

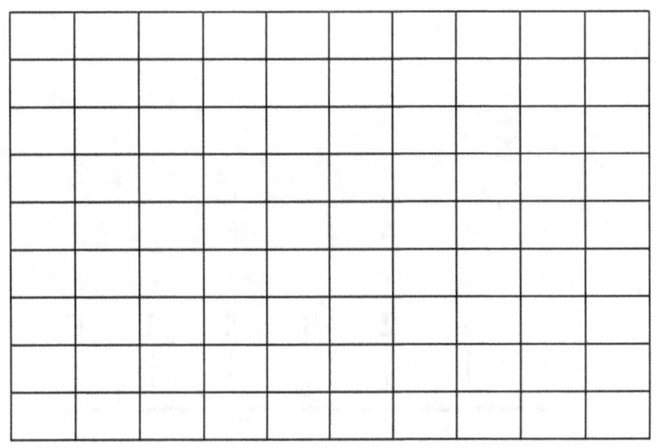

$30 \div 5 =$ ___

Divide by skipping numbers on number line

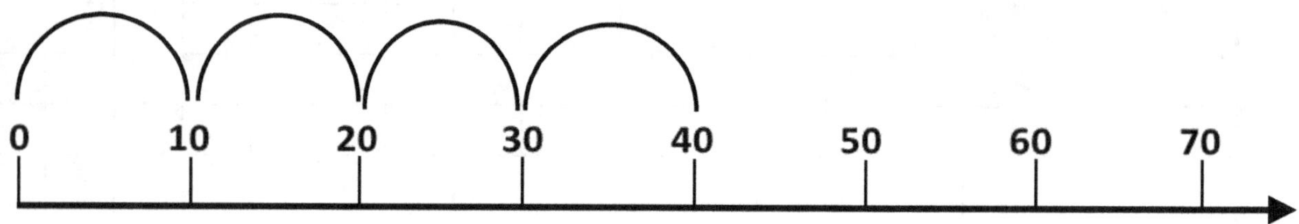

$$40 \div 10 = 4$$

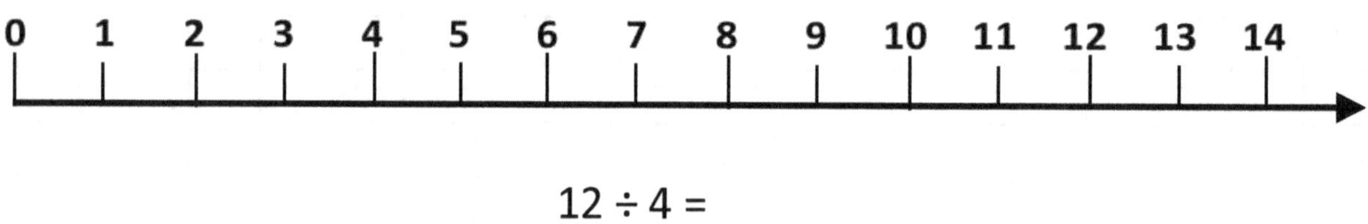

$$12 \div 4 =$$

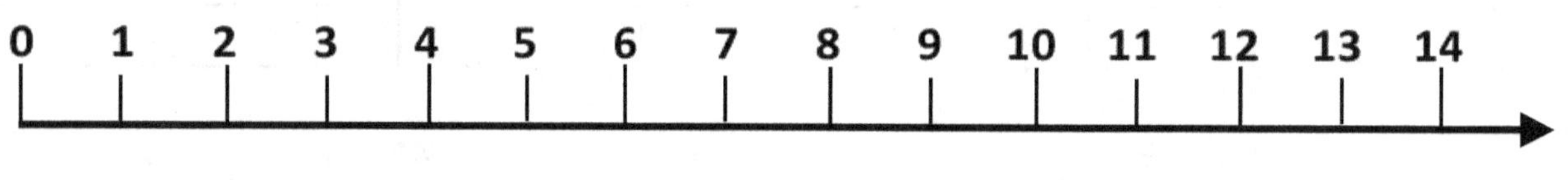

$$12 \div 6 =$$

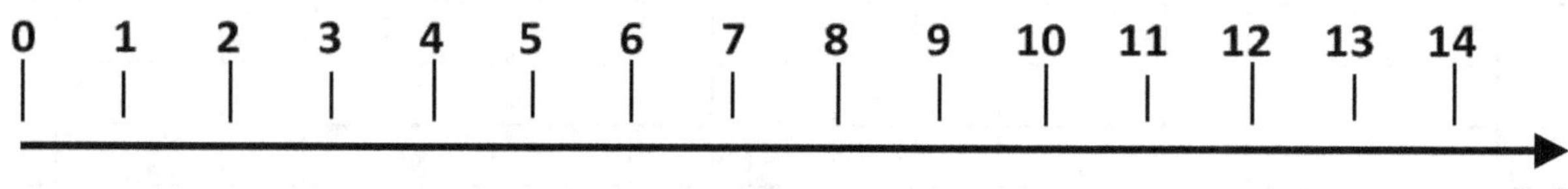

$$10 \div 2 =$$

Divide by skipping numbers on number line

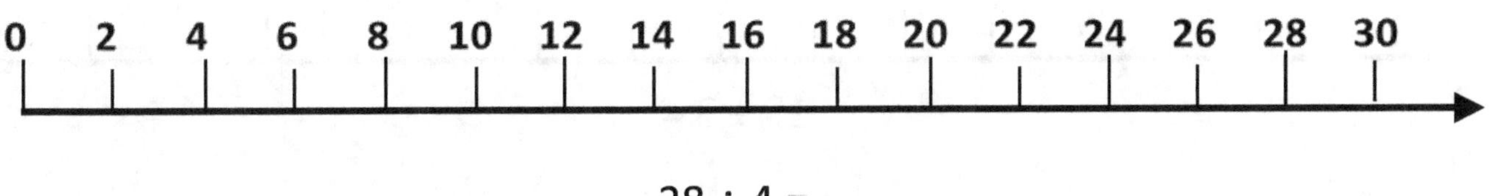

$$28 \div 4 =$$

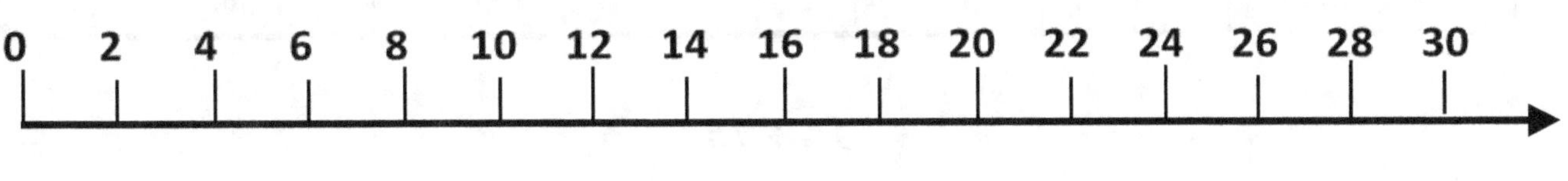

$$24 \div 6 =$$

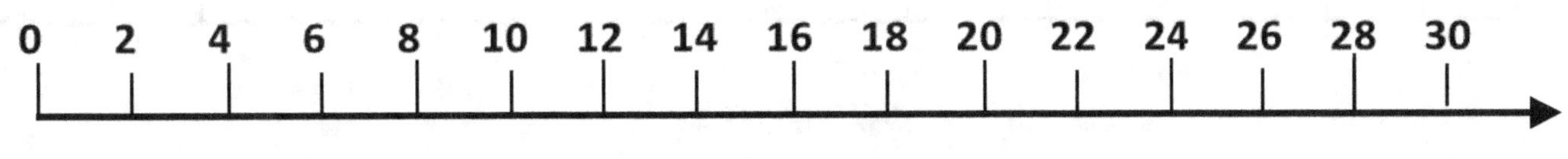

$$30 \div 5 =$$

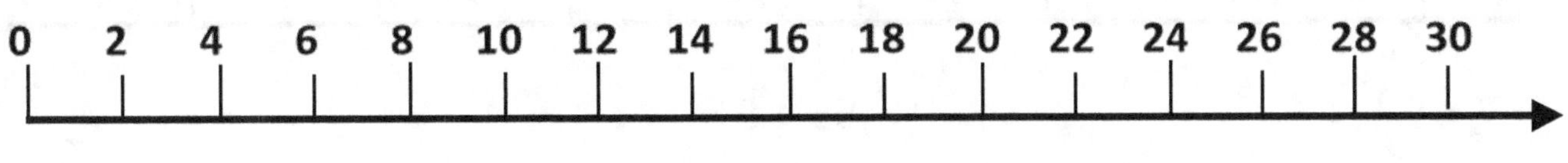

$$25 \div 5 =$$

Divide by skipping numbers on number line

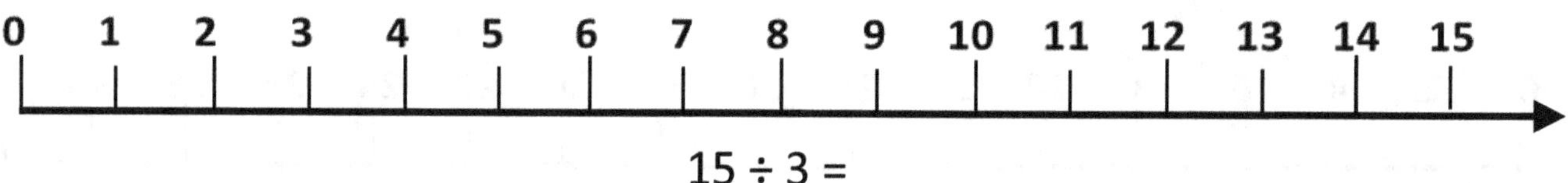

$$15 \div 3 =$$

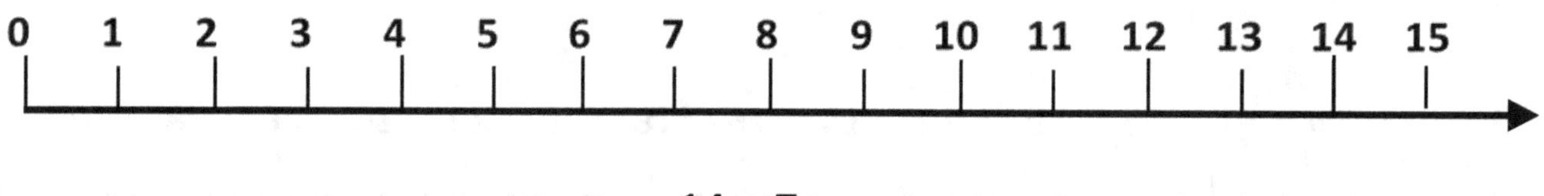

$$14 \div 7 =$$

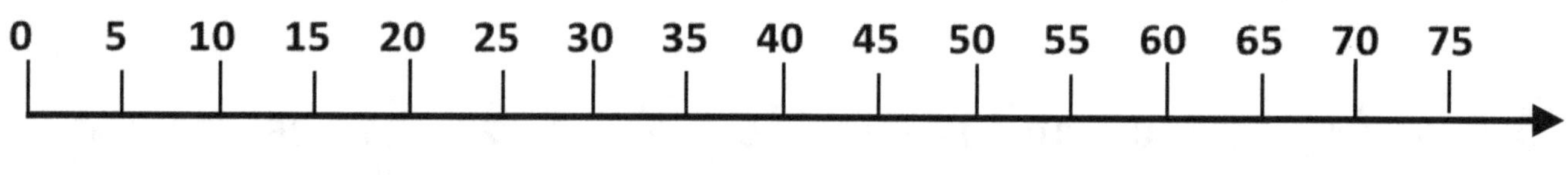

$$70 \div 10 =$$

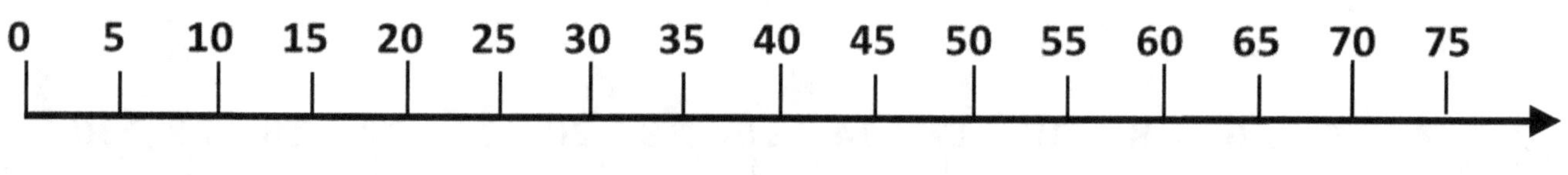

$$65 \div 5 =$$

Divide to complete

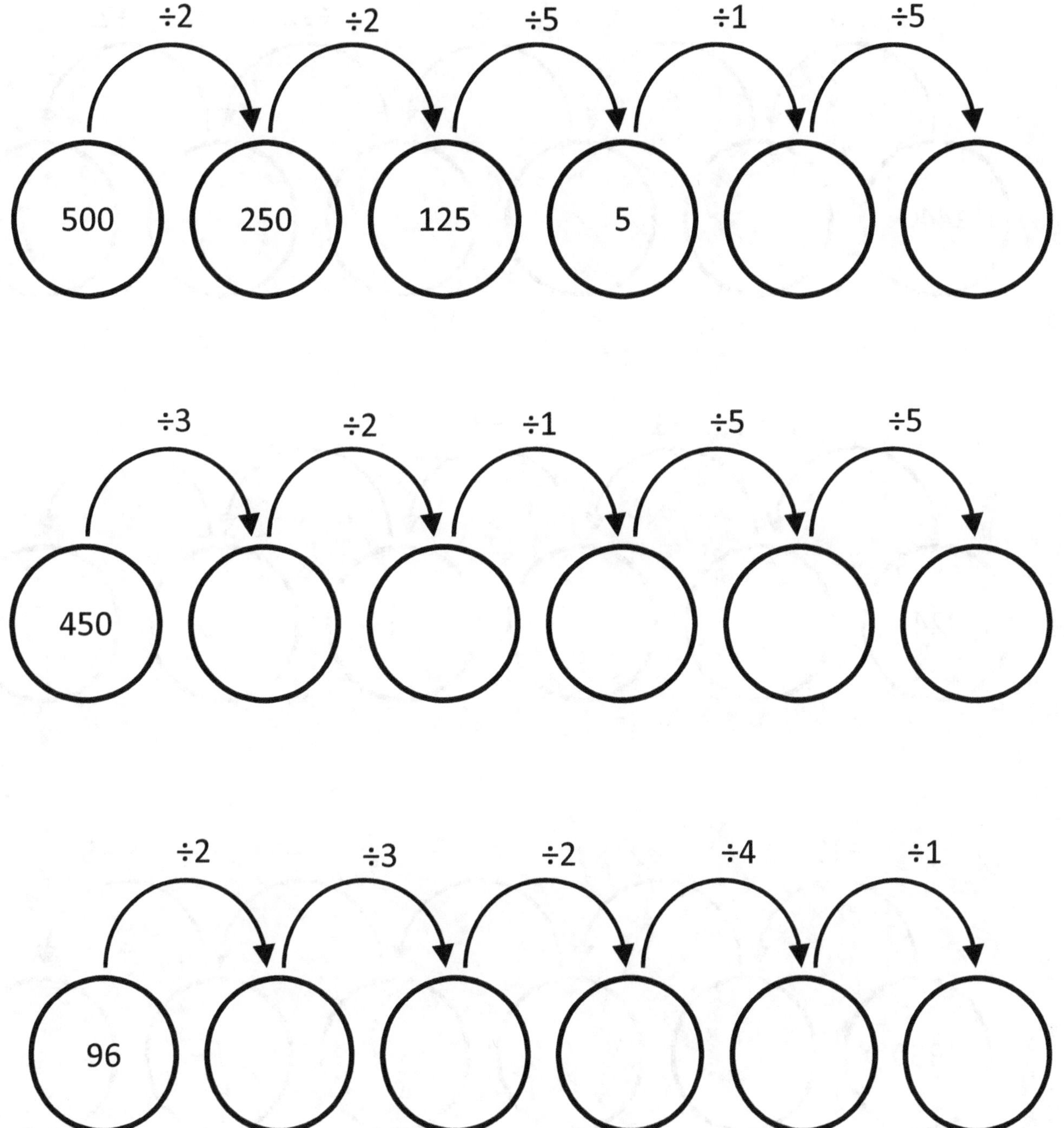

Divide to complete

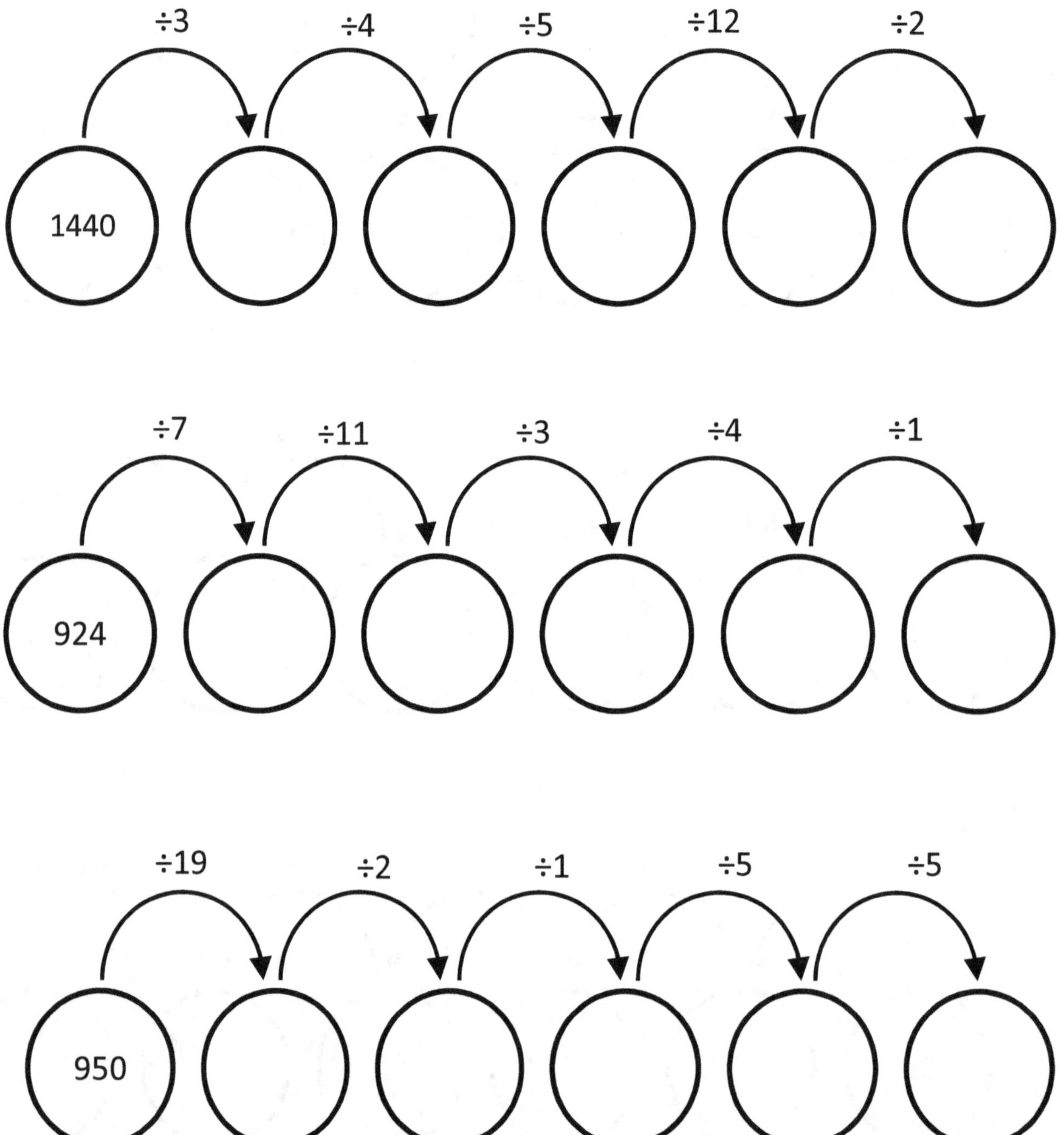

Divide to complete

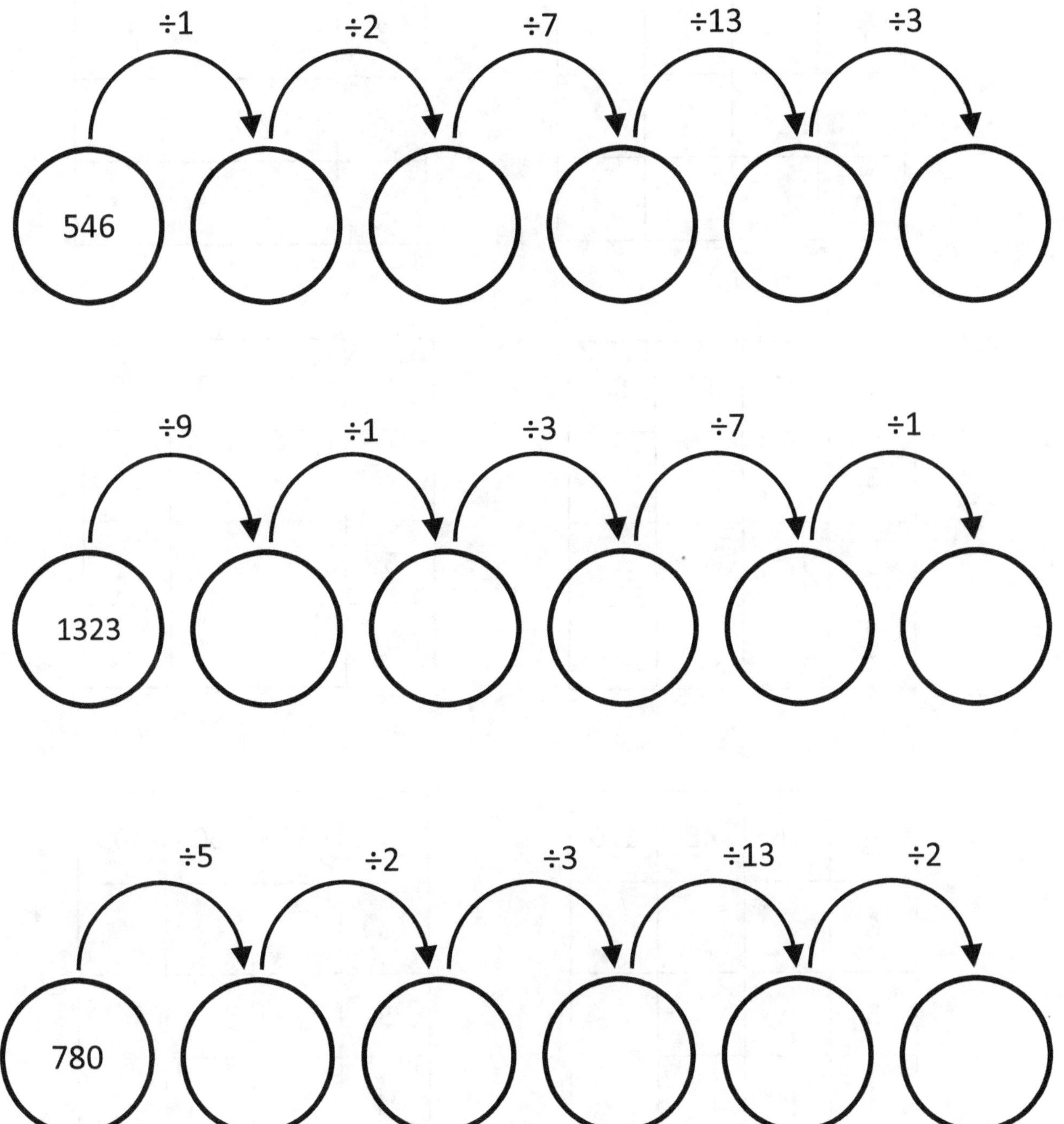

Division Matrix

÷	20	24	36
2			
4			
6			

÷	9	18	36
3			
6			
9			

÷	30	60	90
3			
5			
6			

÷	10	50	70
2			
5			
10			

÷	96	48	120
6			
8			
4			

÷	45	60	105
3			
5			
15			

Division Matrix

÷	60	180	300
5			
10			
12			

÷	36	108	216
6			
12			
18			

÷	120	300	780
10			
20			
30			

÷	108	216	512
3			
9			
12			

÷	96	192	384
4			
12			
16			

÷	110	220	484
1			
11			
22			

Division Matrix

÷	100	500	1000
10			
20			
50			

÷	60	180	360
10			
15			
20			

÷	300	420	660
10			
12			
15			

÷	273	819	1092
3			
13			
21			

÷	105	210	525
5			
7			
15			

÷	132	396	792
2			
12			
22			

Word Problems

1. There are 72 students in a school. They need to be divided into 6 equal size groups. Calculate the number of students in each group.

2. Ms. Lily is packing cupcakes in to boxes. There are 96 cup cakes to be packed in 6 boxes. How many cupcakes will be packed in each box?

3. There are 15 books on each rack in a library. If the library has 180 books in total, how many racks are used to store these books?

4. Mr. Smith is distributing 200 cents among his 4 children. How much will each child get?

5. In a super market, there are 450 packets of snack stored on 15 racks. Calculate the number of packets on each rack.

Word Problems

1. There are 810 students in a school. Each classroom has 30 chairs to sit. How many classrooms are needed in the school to accommodate all students?

2. An ice-cream seller sells an ice-cream for 3 euro. If he made 270 euro in a day, how many ice-cream he has sold?

3. Anna bought 560 crayons. Crayons comes in pack of 20 each. How many packs did Anna bought?

4. It's a soccer season. Total of 131 player came to town. If each team has 11 players, how many total teams will play?

5. Mia arranged 510 marbles in 30 bags. How many marbles are there in each bag?

Word Problems

1. Emma collected 1260 Euros in a month at her store. Assuming she has collected same amount of money every day and the month has 30 days, how much she collected in a day?

2. Sophia has 1026 beads. She started making necklace with beads. At the end she has 19 necklaces. Each necklace hold same number of beads. How many beads are there in each necklace?

3. Ticket to the amusement park is priced at 9 Euros. On a particular day, ticket counter has sold tickets worth 927 euros. Calculate the number of tickets sold on that day.

4. A bakery baked 565 cupcakes on Monday. They packed it in boxes. Each box can hold 24 cup cakes. How many boxes they packed and how many cupcakes left with the bakery?

5. Town planning office brought 1840 saplings for a garden to be planted in rows having equal number of plants. There can be 32 plants in a row. How many rows of plant will be there and how many plants will left out.

Match the following

Left	Right
534 ÷ 89	12
560 ÷ 20	1
720 ÷ 60	59
672 ÷ 56	73
876 ÷ 12	28
826 ÷ 14	13
393 ÷ 3	6
481 ÷ 37	9
62 ÷ 62	6
580 ÷4	18
168 ÷ 28	2
138 ÷ 69	145
864 ÷ 48	131
810 ÷ 90	12

Match the following

10	●
9	●
3	●
9	●
8	●
3	●
81	●
22	●
22	●
2	●
9	●
0	●
3	●
8	●

●	531 ÷ 59
●	946 ÷ 43
●	960 ÷ 96
●	720 ÷ 90
●	891 ÷ 11
●	576 ÷ 64
●	276 ÷ 92
●	190 ÷ 95
●	638 ÷ 29
●	258 ÷ 86
●	57 ÷ 19
●	882 ÷ 98
●	560 ÷ 70
●	0 ÷ 96

Solve the division and then match the division with its other division fact

Left column	Right column
392 ÷ 98 =	184 ÷ 4 =
897 ÷ 39 =	624 ÷ 13 =
267 ÷ 89 =	686 ÷ 49 =
184 ÷ 46 =	340 ÷ 10 =
990 ÷ 33 =	897 ÷ 23 =
624 ÷ 48 =	273 ÷ 7 =
0 ÷ 87 =	39 ÷ 1 =
686 ÷ 14 =	392 ÷ 4 =
360 ÷ 15 =	990 ÷ 30 =
273 ÷ 39 =	468 ÷ 18 =
39 ÷ 39 =	360 ÷ 24 =
468 ÷ 26 =	267 ÷ 3 =
340 ÷ 34 =	760 ÷ 19 =
760 ÷ 40 =	0 ÷ 87 =

Solve the following division problems

Solve the following division problems

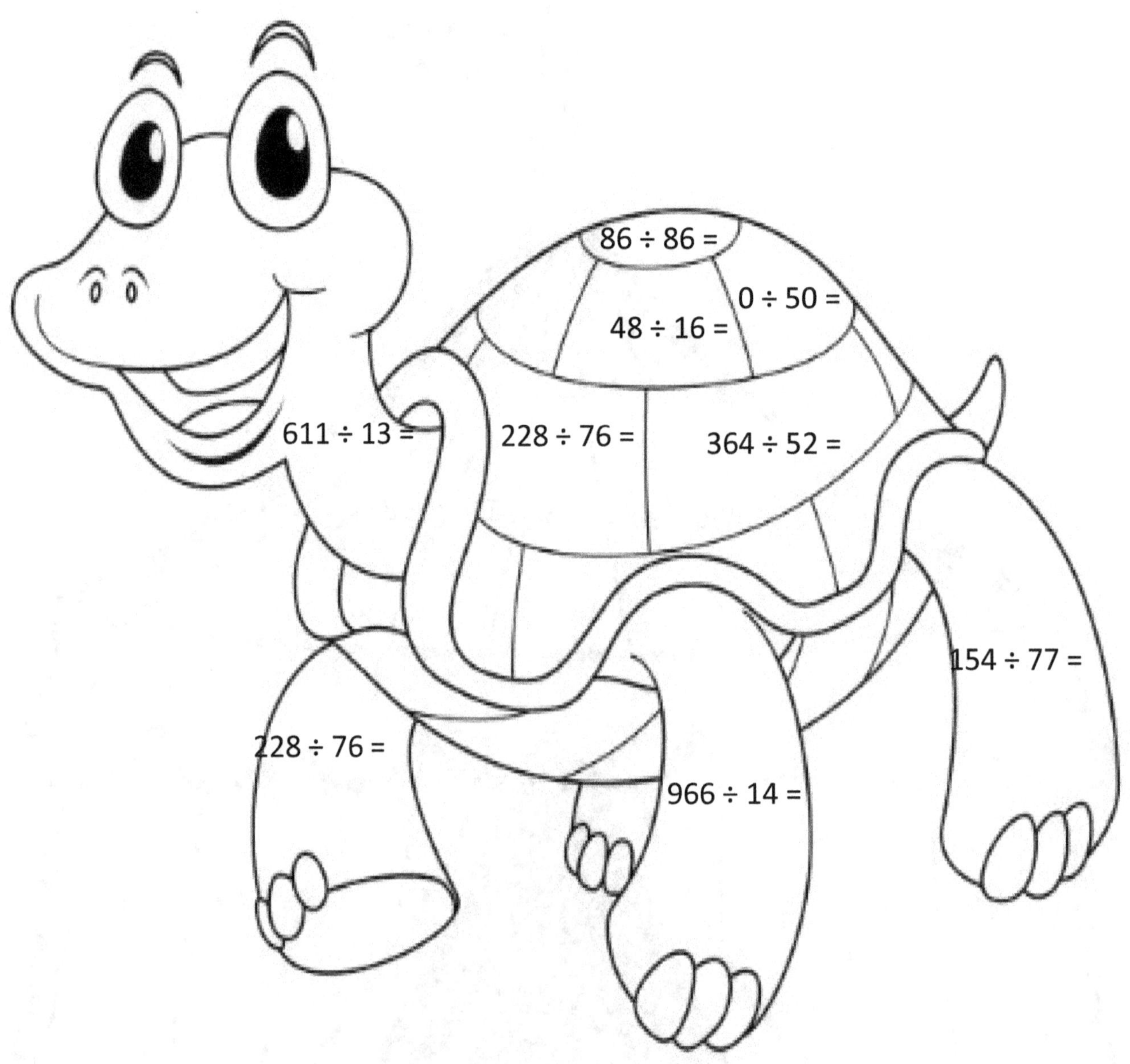

Solve the following division problems

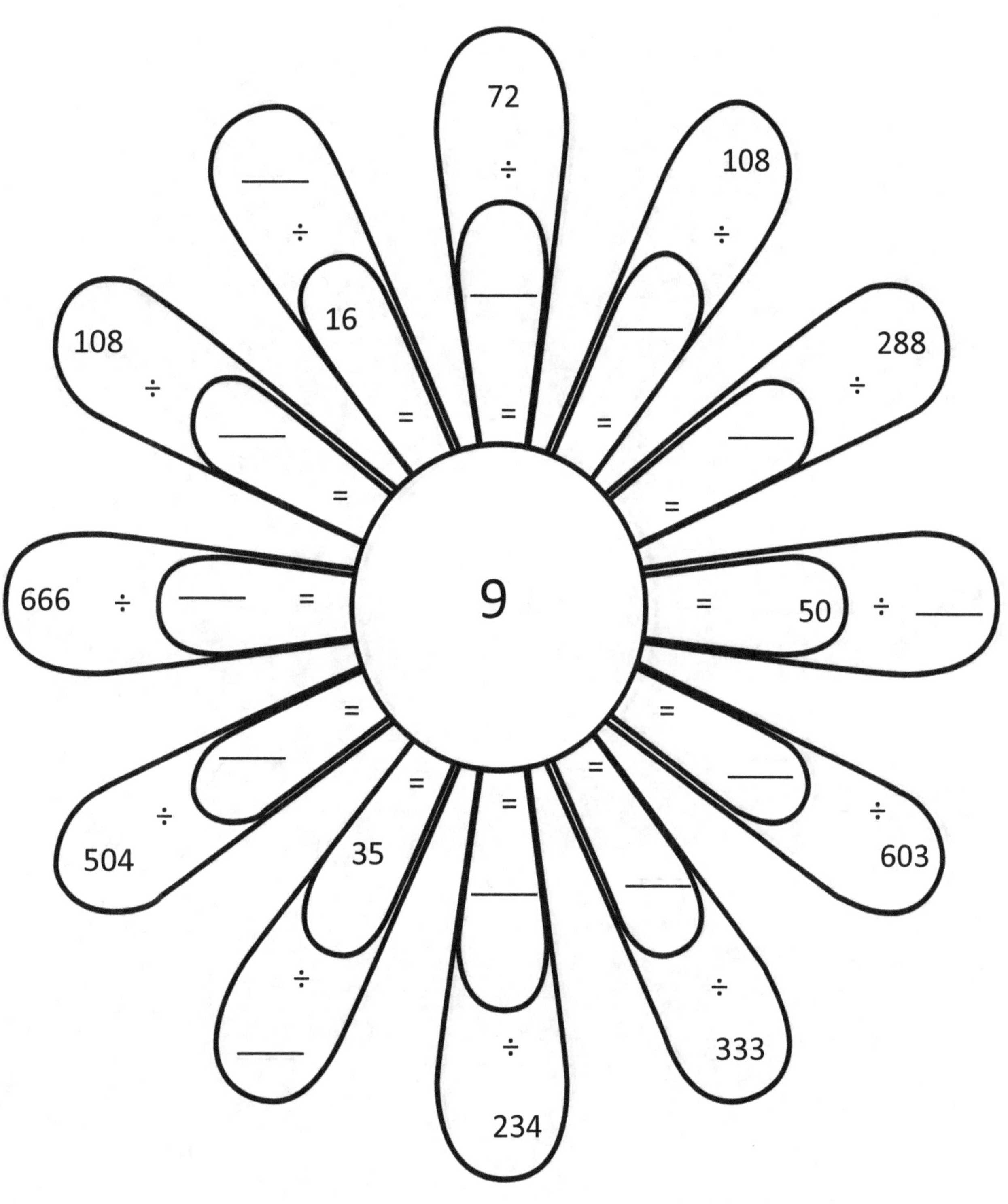

Solve the following division problems

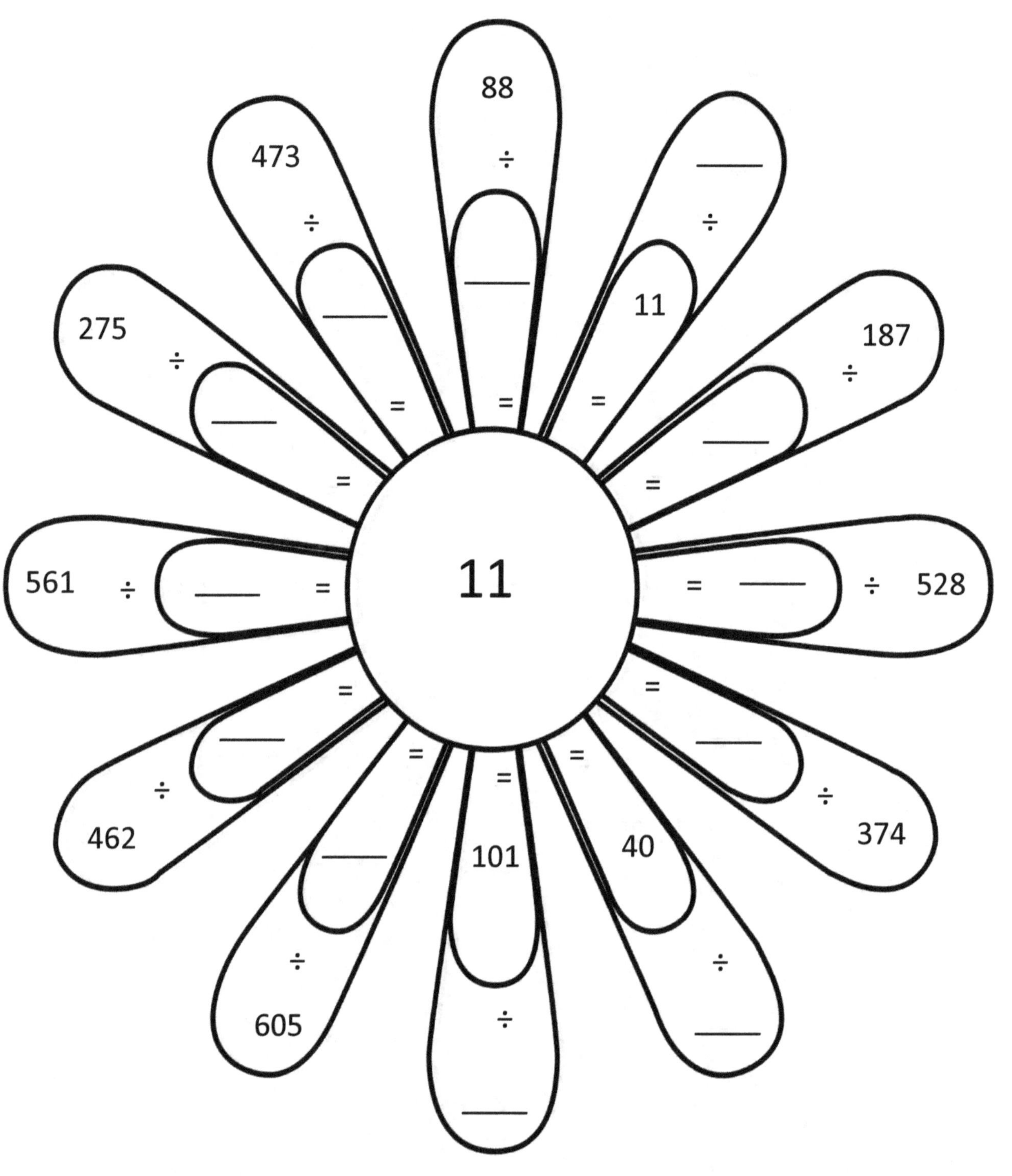

Write down all the division and multiplication facts of given numbers

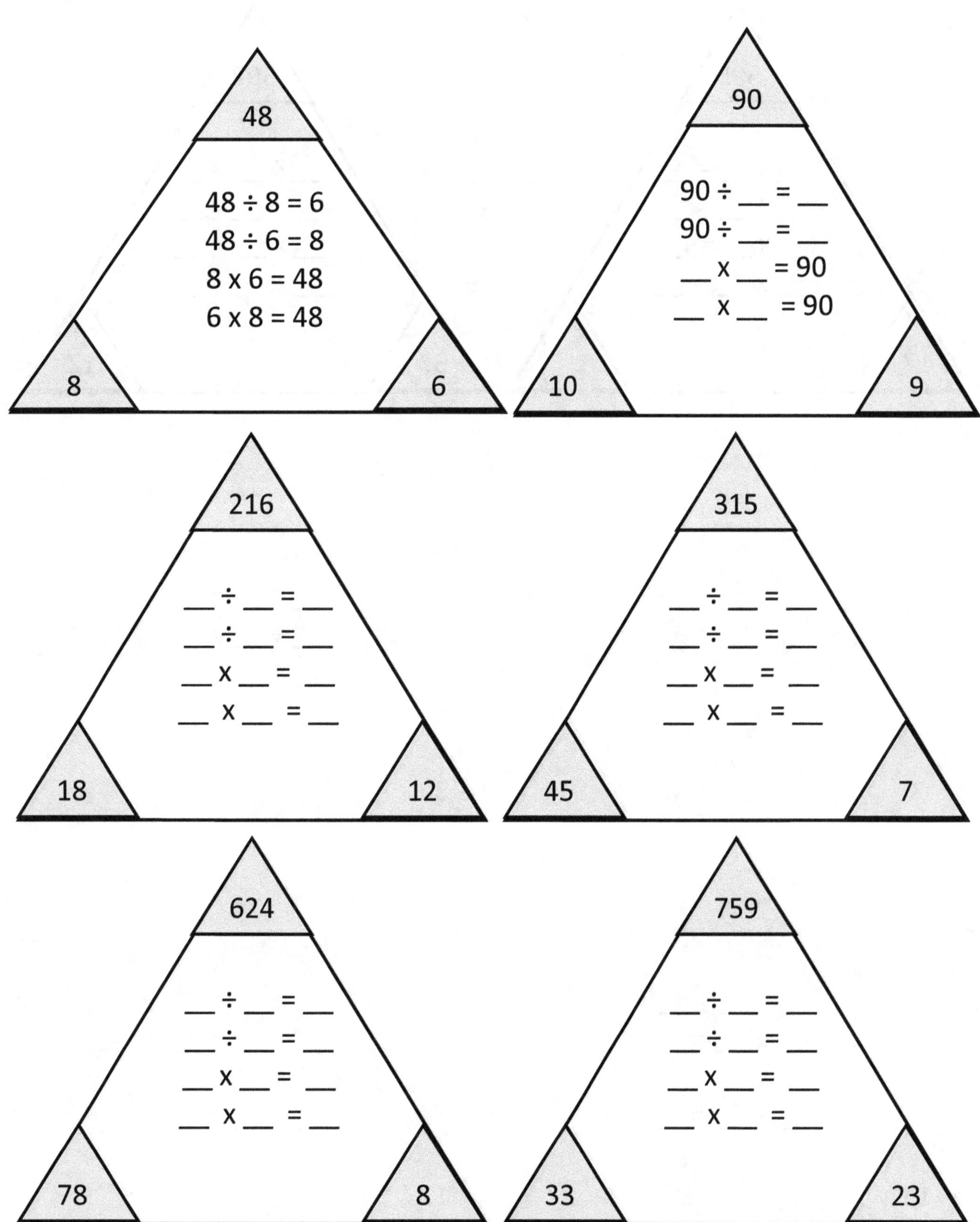

Write down all the division and multiplication facts of given numbers

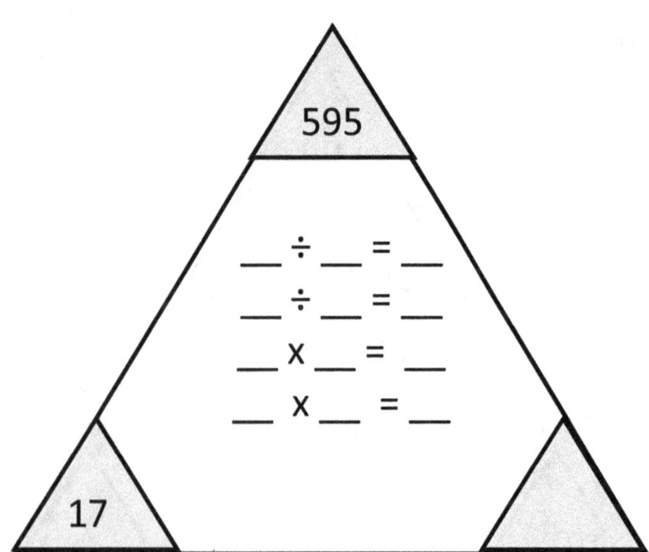

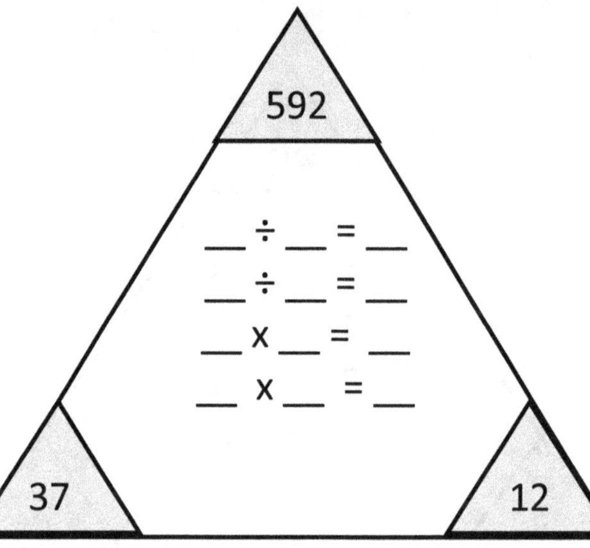

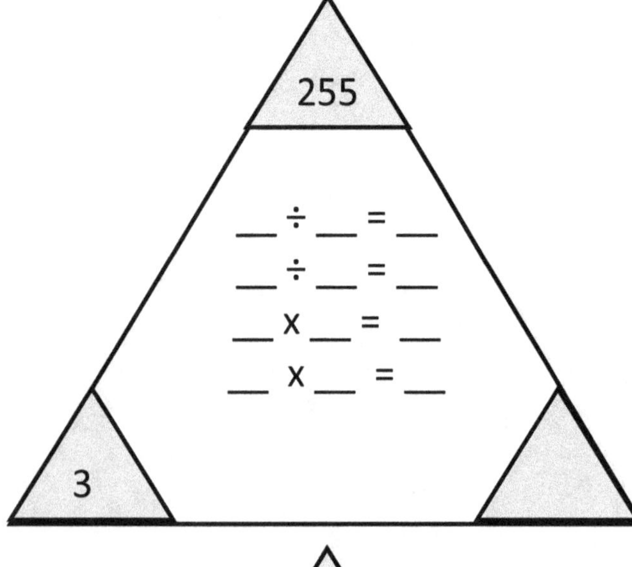

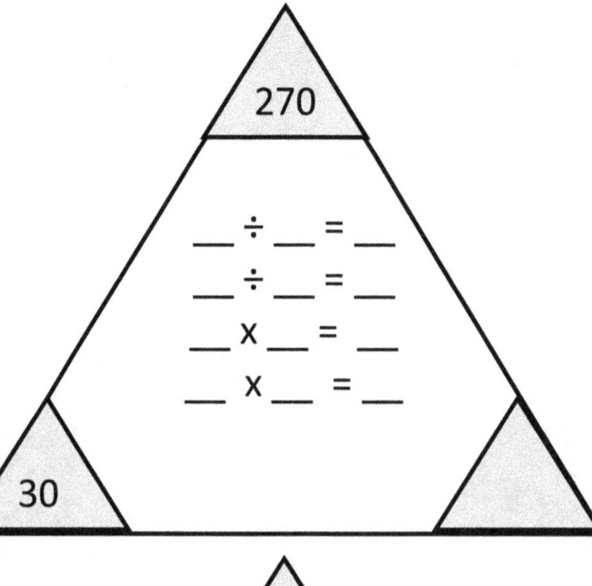

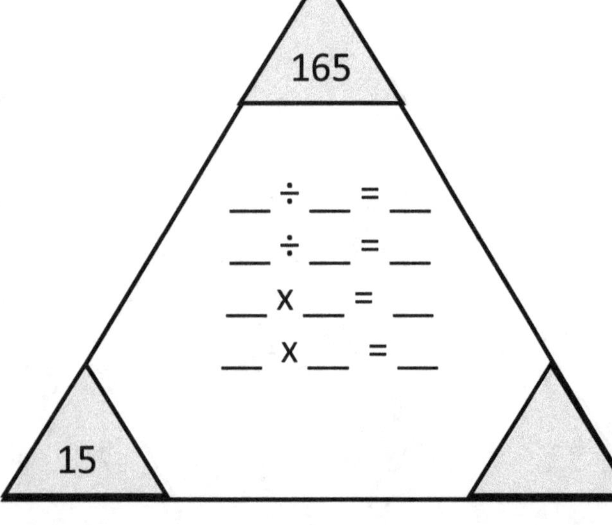

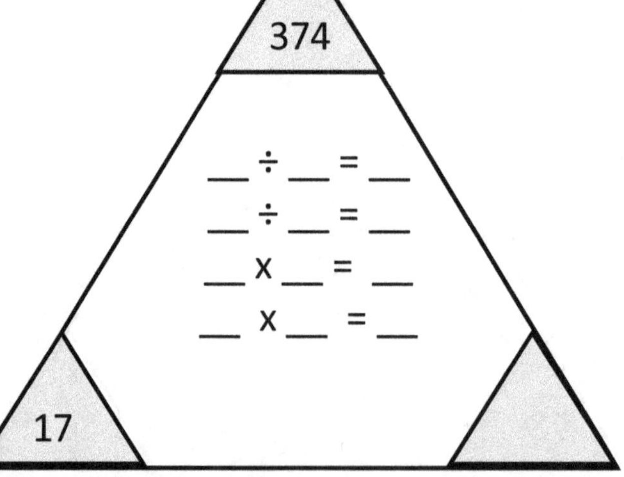

Write down all the division and multiplication facts of given numbers

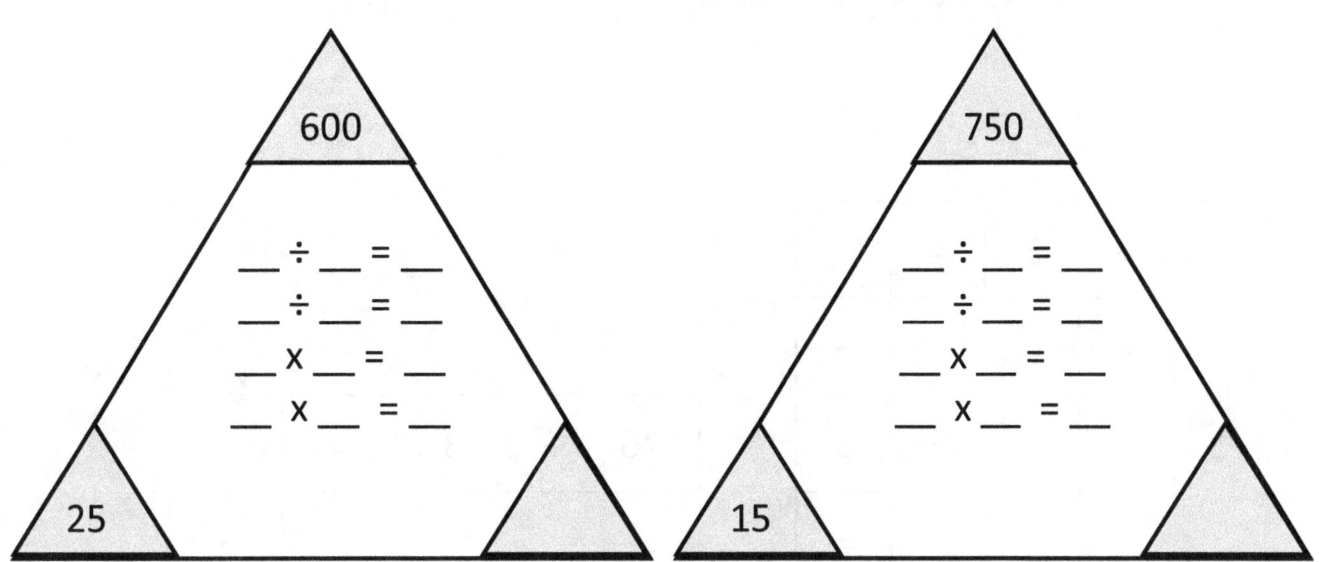

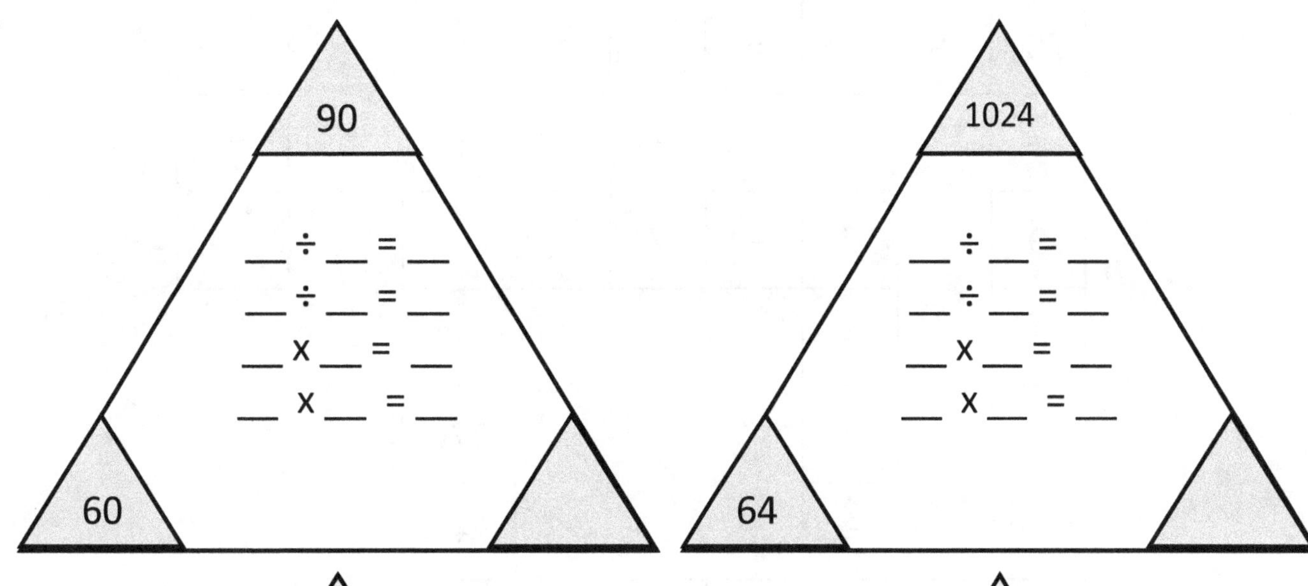

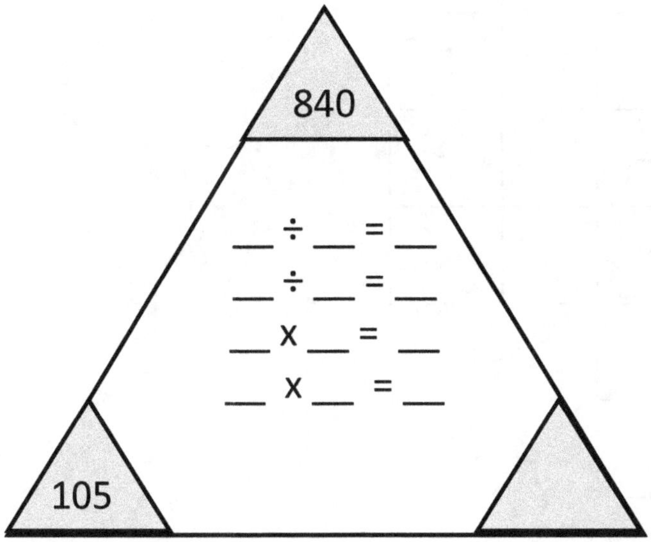

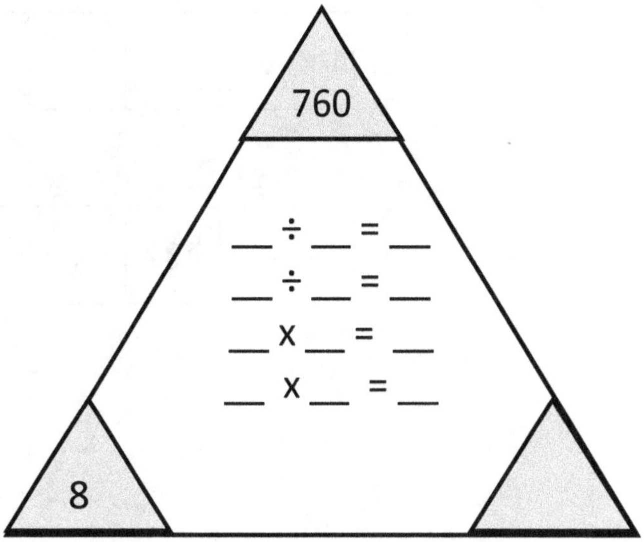

Division Puzzle

	56									
60	÷	3	=							
	4			÷						
	=			5		126	÷	3	=	
				=		÷				÷
		72	÷		=					6
		÷				=				=
		9		133	÷	7	=			
		=		÷						
				19				270		
				=				÷		
		504	÷		=			5		
						÷		=		
				432	÷	8	=			
						=				

Division Puzzle

<table>
<tr><td></td><td></td><td></td><td></td><td></td><td></td><td></td><td></td><td>729</td><td></td><td></td><td></td><td></td></tr>
<tr><td></td><td></td><td></td><td></td><td></td><td></td><td></td><td></td><td>÷</td><td></td><td></td><td></td><td></td></tr>
<tr><td>1944</td><td>÷</td><td>54</td><td>=</td><td></td><td>÷</td><td>4</td><td>=</td><td></td><td></td><td>1323</td><td></td><td></td></tr>
<tr><td></td><td></td><td></td><td></td><td>÷</td><td></td><td></td><td></td><td>=</td><td></td><td>÷</td><td></td><td></td></tr>
<tr><td>72</td><td>÷</td><td></td><td>=</td><td>6</td><td></td><td></td><td></td><td></td><td>÷</td><td>27</td><td>=</td><td>3</td></tr>
<tr><td></td><td></td><td></td><td></td><td>=</td><td></td><td></td><td></td><td></td><td></td><td>=</td><td></td><td></td></tr>
<tr><td></td><td></td><td></td><td></td><td></td><td></td><td></td><td></td><td></td><td></td><td></td><td></td><td></td></tr>
<tr><td></td><td></td><td></td><td></td><td></td><td></td><td></td><td></td><td></td><td></td><td>÷</td><td></td><td></td></tr>
<tr><td></td><td></td><td></td><td></td><td>1156</td><td></td><td></td><td></td><td>1183</td><td>÷</td><td>7</td><td>=</td><td></td></tr>
<tr><td></td><td></td><td></td><td></td><td>÷</td><td></td><td></td><td></td><td>÷</td><td></td><td>=</td><td></td><td></td></tr>
<tr><td></td><td></td><td>1547</td><td>÷</td><td>17</td><td>=</td><td></td><td>÷</td><td>13</td><td>=</td><td></td><td></td><td></td></tr>
<tr><td></td><td></td><td>÷</td><td></td><td>=</td><td></td><td></td><td></td><td>=</td><td></td><td></td><td></td><td></td></tr>
<tr><td></td><td></td><td>13</td><td></td><td></td><td></td><td></td><td></td><td></td><td></td><td></td><td></td><td></td></tr>
<tr><td></td><td></td><td>=</td><td></td><td>÷</td><td></td><td></td><td></td><td>÷</td><td></td><td></td><td></td><td></td></tr>
<tr><td></td><td></td><td></td><td>÷</td><td>17</td><td>=</td><td></td><td>÷</td><td>7</td><td>=</td><td></td><td></td><td></td></tr>
<tr><td></td><td></td><td></td><td></td><td>=</td><td></td><td></td><td></td><td>=</td><td></td><td></td><td></td><td></td></tr>
<tr><td></td><td></td><td></td><td></td><td></td><td></td><td></td><td></td><td></td><td></td><td></td><td></td><td></td></tr>
</table>

Division Puzzle

				2623								
				÷								
1075	÷	25	=			3360						2780
÷				=		÷						÷
215		2440	÷		=		÷			=		5
=						=						=
		3192	÷		=							
		÷					÷					÷
		152					2					4
		=					=					=
		1995	÷		=							
		÷										
987	÷	7	=		÷		=	47	÷	47	=	
		=										
			÷	15	=		÷	1	=			

www.ingramcontent.com/pod-product-compliance
Lightning Source LLC
Chambersburg PA
CBHW080310030726
47593CB00009B/2709